Farming Machinery - Combine Harvesters

With Information on the Operation and Mechanics of the Combine Harvester

By

Various Authors

Contents

Combine Harvesters

"I had the pleasure of seeing this wonderful machine at work in California in 1887. It was propelled by 16 mules, harnessed behind, so as not to be in the way; but steam-power is now used."

The Wonderful Century. Alfred Russel Wallace, 1898.

The idea of harvesting in one operation is now more than a hundred years old. Two separate combined harvesting and threshing machines were patented in the United States in 1836, but neither achieved any practical success. One of the earliest practical machines was a stripper type combine invented in Australia in 1845. Little advance was made until about 1860, when machines began to be rapidly developed in California, where conditions are especially favourable to the combined operation. Improvements in both reapers and threshers were incorporated in the combined machines, and by 1890 combines were not at all uncommon in California.

It was not until the period just after the 1914-18 war that the combine began to spread appreciably to the east of the Rocky Mountains into the semi-arid sections of the United States and the Canadian Prairies, and it was only after about 1930 that small machines began to be generally used in the mixed farming areas of the Middle West and the Eastern States. The combine is essentially a labour-saving machine, and the main reason for its spread throughout North America has been the economic necessity of reducing production costs by reducing man-labour.

The combine was introduced into this country in 1928. In 1930, four privately owned machines were at work here; in 1931 the number had risen to ten, and the next year it was doubled. There were just over fifty machines at work in England during the 1937 harvest. Shortage of labour during the war years greatly accelerated adoption of combining. About 500 machines were in use in 1941 and nearly 1,000 in 1942. By 1947, when about 5,000 machines were in use, the practice of harvesting by

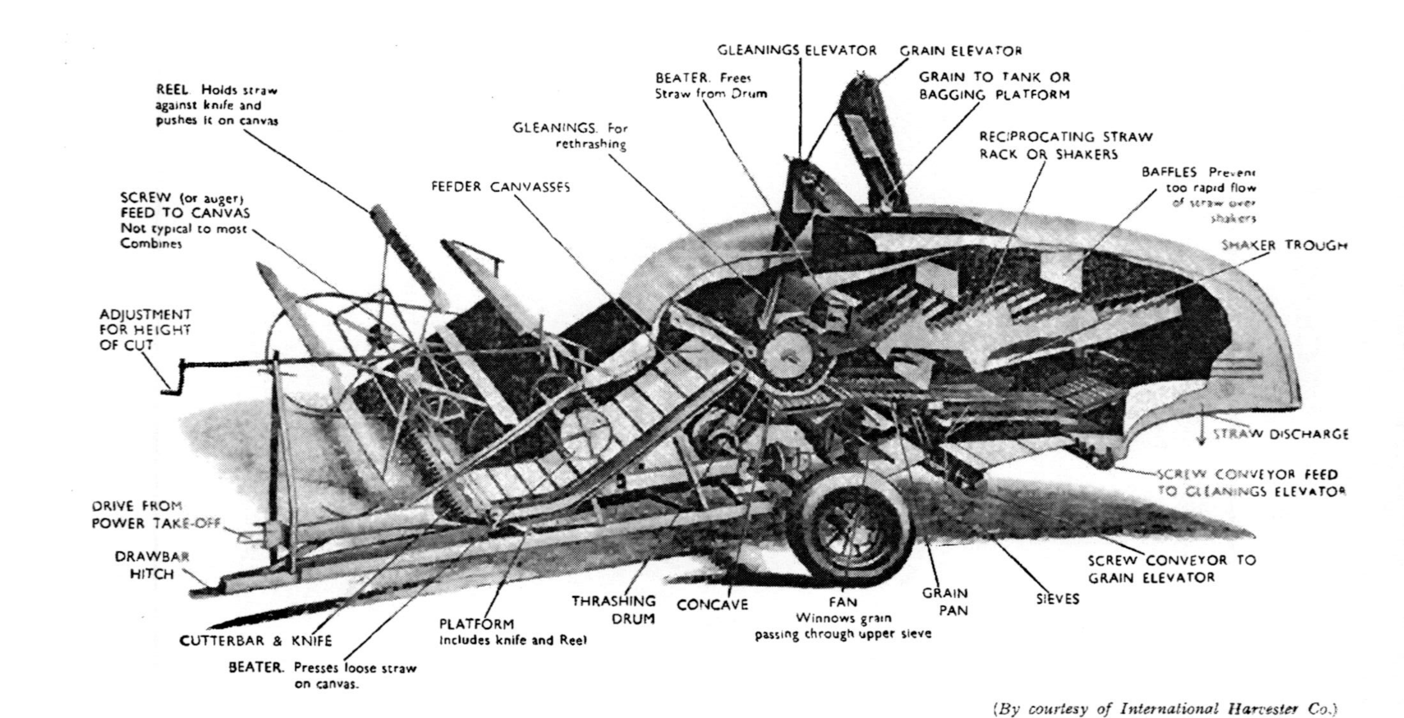

(*By courtesy of International Harvester Co.*)

FIG. 185.—SMALL TRAILED P.T.O.-DRIVEN COMBINE HARVESTER IN SECTION

combine had spread to all important grain-producing areas in Britain, and since that time the numbers used have increased substantially, about 20,000 being on British farms by 1952.

Methods of Harvesting by Combine. Though combine harvesters are mostly used to cut and thresh in one operation, they may also often be used to advantage to pick up and thresh crops which have previously been cut and left in the swath, or as stationary or portable threshers for dealing with crops which have been stooked or stacked.

WINDROWING. The technique of "windrowing" or "swathing" has attracted considerable interest on account of the possibility that it might assist the combining of grain of low moisture content and eliminate the need for drying. In fact, however, substantial drying of corn in the windrow does not take place except in weather when standing crops dry out well, and it is for other reasons that windrowing of corn is worth consideration. It provides a rapid means of saving crops from shedding if too many fields ripen at the same time. It is also a useful technique when weeds are present in the crops, for mixed corn or any crop which ripens unevenly, or occasionally for crops in which clover has grown too high. Peas differ from corn in that the crop often needs to be windrowed in order to hasten drying.

When windrowing is practised the corn may be cut either by a special windrower or swather, or by a binder from which the compressor arm has been removed. Figs. 186 and 187 show two types of windrowing machines. The trailer type is designed primarily for corn, and is similar to the machines now widely used throughout the Canadian Prairies. The 10 ft. cut makes a good swath, and allows plenty of room for delivering the swath on to stubble which has not been run over by the wheels of either tractor or windrower. The height of cut is adjustable from about 3 in. to 18 in. A pick-up reel can be fitted. A clutch is provided to stop the conveyor at corners. It is a one-man machine, and can windrow a standing crop at a high speed.

The tractor-mounted type of swather was originally designed for harvesting crops such as peas, so has different characteristics.

This machine is designed for close cutting and for dealing with crops which require lifting. It is usually equipped with a pick-up reel. Both front-mounted and rear-mounted machines are available.

FIG. 186.—10-FT.-CUT TRAILED P.T.O.-DRIVEN WINDROWER. (HARRISON MCGREGOR AND GUEST.)

FIG. 187.—FRONT-MOUNTED SWATHER UNDERGOING OUT-OF-SEASON TESTING. (LEVERTON.)

In general, the technique of windrowing can only be recommended where the corn is standing, and has a reasonably sturdy stubble which will support the crop well clear of the ground. As a general rule a stubble 6-8 in. high is needed, and laid crops are unsuitable for windrowing. On the other hand, provided that the combine follows fairly close behind the swather, a type of tractor-mounted machine which can almost shave the ground has been proved invaluable for cutting and windrowing crops which are laid so flat that no combine harvester cutter-bar can get under the straw. Tractor-mounted swathers are now widely used for harvesting peas, both green for delivery to green pea threshers and also ripe peas for seed and packeting.

When a binder is used for windrowing, great care must be taken to ensure that the swath does not fall into a wheel-mark, and it may be necessary to offset the tractor.

When seed crops are windrowed and heavy rain beats them

FIG. 188.—WINDROW AERATOR. (TASKERS.)

down, there are few machines suitable for raising the windrows so that they can dry out again. Most of the swath turners used for haymaking are quite unsuitable. They knock out seed and disturb the swath so that it does not feed evenly into the combine when it is picked up later.

The WINDROW AERATOR is a machine specially designed for lifting such crops. It consists of a pick-up attachment, usually of the draper type, which gently raises the windrow and delivers it over the rear of the machine. Some farmers who practise windrowing consider such a machine almost indispensable.

A combine harvester which is used for picking up from the windrow must, of course, be equipped with a pick-up attachment in place of the cutter-bar.

Main Types of Combine. Practically all combines consist essentially of the cutting mechanism of a binder attached to a travelling thresher. An exception is the McConnell-Wild Harvester, which threshes the standing crop and leaves the cutting and collection of the straw to be dealt with as a separate operation. Excluding such machines, there are several quite distinct types of combine. The simplest type is the small machine of up to 6 ft. cut, mounted on two fixed wheels, and with cutter bar, elevator, threshing drum and shakers arranged so that the straw travels straight back through the machine. A modification of this is a simple type in which the corn goes straight back from cutter-bar to drum, and then on to shakers disposed at right angles to the direction of travel.

Larger machines generally have the cutter-bar off-set to the right, and carried either by a wheel on a sliding axle which can be adjusted according to whether the cutter-bar is attached or folded up ; or by a separate detachable wheel. With this type the corn has to be carried to the left, as on a binder platform, and then passes back to the drum. Some combines employ canvases similar to those on the binder, but others have a large Archimedean screw or auger, which works well in most conditions, especially with short straw. With long straw, rubberized canvases may be better.

The other main type of combine is the self-propelled machine, with four wheels, cutter-bar at the front, and canvases or auger running towards the centre to deliver the corn to a centrally disposed drum. Some features of the construction of the various types are briefly described below. The great advantage of the self-propelled machine, apart from the fact that a tractor is saved in a busy time, is that it is able to go straight into the crop without any opening out, and can cut it in the most convenient direction.

It also has a large number of gears, and this enables it to deal effectively with a wide range of crops. The driver has a full view of the cutter-bar, and is able to leave laid or otherwise difficult patches if the weather is damp, or to get on with the difficult patches and leave the other when the crop is really dry. Some combines are constructed with a view to bulk handling of the grain, while others are equipped for bagging.

Power Required. The cutting and threshing mechanism is usually driven by an auxiliary engine mounted on the combine,

FIG. 189.—TWELVE-FOOT CUT SELF-PROPELLED COMBINE WITH GRAIN TANK, HARVESTING WHEAT. (MASSEY-HARRIS.)

but on small machines up to 5-7 ft. cut it may be driven from the tractor power take-off.

The tractor for driving a P.T.O. machine requires adequate power and a good selection of gears at low speeds if it is to deal efficiently with a range of heavy crops. The governor must be efficient, so that difficult conditions do not cause speed to drop unduly. Combining is one of the operations for which a P.T.O. independent of the main drive to the tractor wheels is a great advantage. A tractor of not less than about 30 b.h.p. is needed to provide adequate power for a P.T.O. driven 6-ft. combine in average conditions, but a small tractor may be adequate on flat firm land for one which has an independent engine drive.

Capacity of Combine Harvesters. Experience has shown the difficulty of giving a figure for combine harvester capacity which is of general use in Britain's widely varying conditions. In the Southern and Eastern Counties, if a drier is available when necessary, a combine can usually be relied on to deal with 25 acres for each foot of cutter-bar width. Such a figure may, however, be much too high for the wetter conditions often prevalent in the North and West, where a figure of 15-20 acres per foot is a more realistic estimate. On the other hand, there are many farms in the more favoured Southern and Eastern districts where 30 acres per foot is a normal figure, and where 40 acres has often been exceeded.

The average amount cut per machine in Britain is steadily declining as combines are used on smaller farms. A decline in average annual utilization is not necessarily altogether wasteful. It provides a reserve of combining capacity for use in unfavourable seasons, and enables the combine to be used only when conditions are really suitable, and the grain needs little drying. Little-used combines, if well maintained and carefully stored when not in use, should have a long life.

The Cutting Mechanism. The platform can be adjusted to leave various lengths of stubble, the range frequently being from 3 in. to over 2 ft. The position of the reel may be easily adjusted when the machine is at rest, but is not usually moved during work, as it is with the binder. A knife with serrated teeth is employed on some machines. Such knives will work for long periods without sharpening, and are seen at their best when cutting dry, upstanding corn ; but they compare unfavourably with knives fitted with plain sections when cutting corn containing a high proportion of weeds or with undersown " seeds ". The cutter-bar and platform are usually balanced so as to allow easy adjustment. Large machines often have a power-operated adjustment for length of stubble.

PICK-UP REEL. For use in laid crops a pick-up reel may be fitted in place of the standard type. The hanging tines are maintained at an adjustable angle by an eccentric mechanism. The reel works best when the crop is laid across the direction of travel.

The Threshing Mechanism does not differ in most essentials from that of the stationary thresher. The feed to the drum is often assisted by a rapidly revolving " beater " but some machines

FIG. 190.—SELF-PROPELLED COMBINE HARVESTER FITTED WITH PICK-UP REEL FOR LAID CORN, AND BAGGING ATTACHMENT. (MASSEY-HARRIS.)

have an upper feeder canvas for this purpose. There is usually a revolving beater just behind the drum, and its purpose is to free straw from the drum and open it out on its way to the shakers. There are three main types of drum and concave, viz. (1) the beater bar type as used on British threshing machines (Fig. 179) ; (2) the peg type, which chops up the straw and absorbs more power (Fig. 181) ; and (3) the flail type, a light variant of the beater type, which knocks the corn out rather than rubbing it out (Fig. 191). The flail type may be rubber-faced. It runs at high speed and is used only on small machines.

Some of the larger machines have shakers similar to those on a thresher, but small machines often have a single wide shaker or straw-rack.

Grain from the shakers and from the concave passes to two or three sieves which reciprocate together. The upper sieve is generally of the louvred type, with serrated adjustable louvres. An air blast winnows out the chaff, and the grain passes through to a second sieve which may be either louvred or with round holes. Grain passing through this sieve goes to the grain elevator. The material off the top of the sieve passes to the tailings or gleaning elevator, which delivers unthreshed ears or pieces of ear back to the drum again.

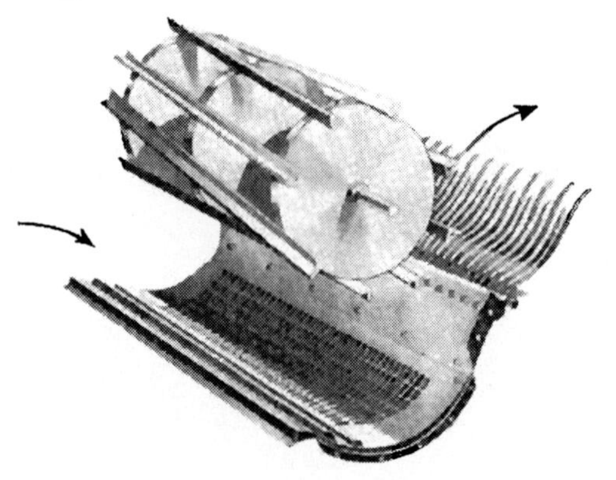

FIG. 191.—FLAIL TYPE DRUM AND CONCAVE

The passage of the corn through a modern self-propelled combine may be followed by reference to Fig. 192. The corn is carried by the auger to the centre of the table, where it is picked up by a chain and slat elevator and raised to the drum. The feeding beater helps it on, and threshing takes place between drum and concave. Grain, chaff and straw pass on to the shakers, and grain return chutes bring back chaff and grain to join that which passed straight through the concave on to the reciprocating grain pan. The mixture of chobs, chaff and grain passes on to the sieves, where an adjustable blast blows the chaff out of the back of the machine. A return pan brings the grain to the grain auger and it passes up the grain elevator to the grain tank or bagging device. Pieces of unthreshed heads which are not blown out with the chaff and will not pass through the sieves are returned by an auger to the tailings elevator, which delivers them again to the drum.

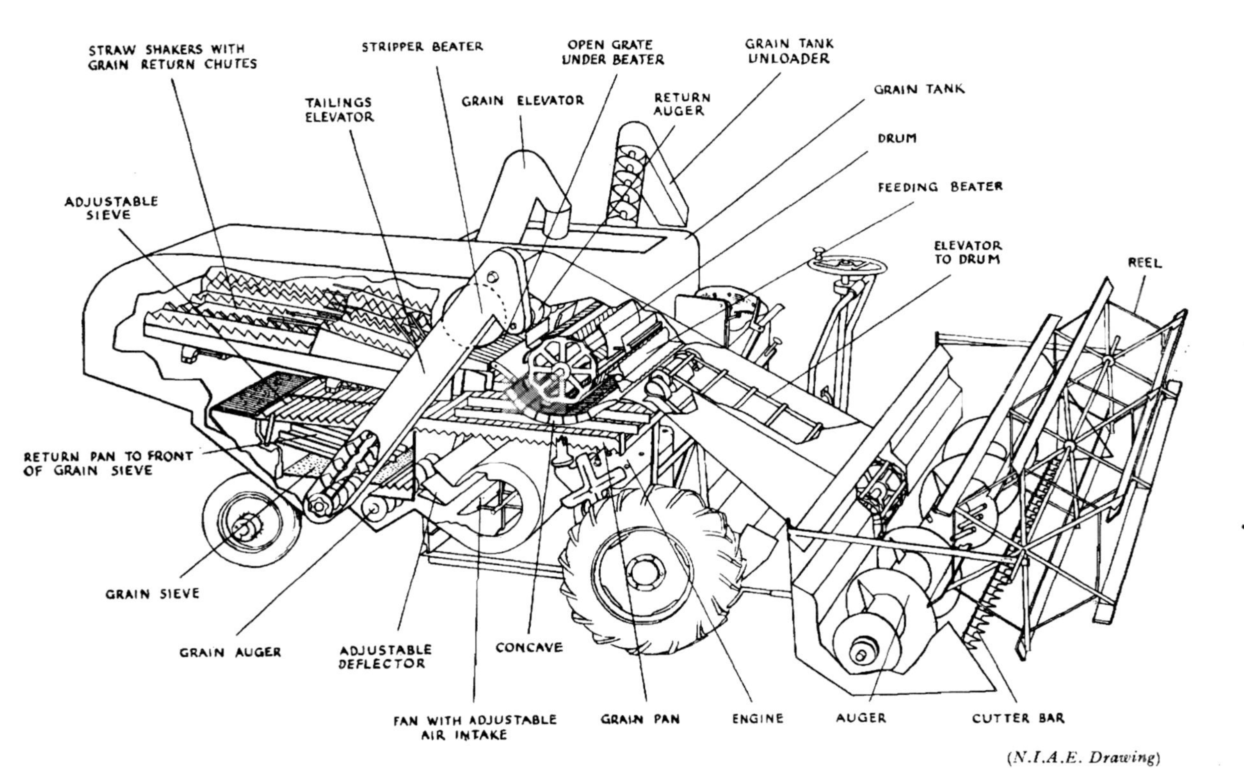

(N.I.A.E. Drawing)

FIG. 192.—SECTION OF A SELF-PROPELLED COMBINE HARVESTER.

Combines do not normally grade the grain. Weed cleaners which take out small seeds for separate bagging are often fitted, but light weed seeds are blown out with the chaff.

The modern combine will work efficiently on fairly steep slopes. In U.S.A., for situations where there are very steep slopes, a special type of machine, called the " hillside " combine, can be employed. This contains adjustments which enable the threshing mechanism to be kept level while the machine is working on side gradients which would be impossible to cultivate without special equipment.

GRAIN HANDLING. The grain may be delivered either into bags or into a grain storage tank mounted on the machine. The latter method is generally more satisfactory if central buildings, well equipped for bulk handling, are available. In ideal conditions the grain is delivered from the grain tank into a tipping motor lorry or trailer which can deliver its load into an elevator hopper at the buildings without the necessity of employing any hand labour. With fields up to two miles from the drier a 2-ton lorry can serve two 12-ft. combines.

Where a tractor and trailer are used for bulk grain transport it is usually convenient to have two trailers, one of which can be left at a corner, handy for the combine.

Bagging the grain is a simple method but it needs an extra man on the combine and may lead to a good deal of work in picking up bags of grain afterwards. Advantages are that no specialized equipment is needed, the picking up can be done when the combine is not at work, and bags known to contain damp or green corn can be kept separate. When a bagging combine is used it is generally a convenience to arrange either to dump the bags in windrows or to transfer them straight on to a trailer. Damp grain in small hessian bags can remain in the open overnight without much risk of damage by heating, but in warm weather, damp grain put into big heaps and covered by a tarpaulin may heat in a few hours.

Operation and Adjustments. It is not usually possible to obtain a perfect sample of grain with the combine, nor is it desirable to overload the cleaning mechanism by attempting to do so. It is important that the drum and shakers should be large enough, in proportion to the size of the cutter-bar, to make possible efficient separation of the grain from the straw. But the threshed grain almost inevitably contains pieces of herbage that cannot be effectively removed while wet, and it is better to leave most of the cleaning to be done by barn machinery rather than to hinder the progress of the harvest.

Where both combine and binder are to be used, it is advisable to use the combine on the lighter and cleaner crops, and on those

FIG. 193.—TWELVE-FOOT SELF-PROPELLED COMBINE HARVESTER USING WINDROW PICK-UP ATTACHMENT. (MASSEY-HARRIS.)

which are badly laid. Wheat and barley should be combined in preference to oats. It should be remembered that the rate of combining is governed by the volume of straw handled, rather than by the weight of grain. Weeds, damp weather and damp straw all reduce output. Thistles and other juicy weeds clog the screens and shakers, and it may be necessary to stop and clean them frequently when operating conditions are bad.

All grain should be dead ripe before being combined. Wheat, for example, generally needs ten days to a fortnight after it is fit for the binder. This longer growing period results in a higher yield and less tail corn. There is, naturally, a risk of oats blowing out and of barley necking when the crops are left to ripen. Clover may be directly combined on a fine, dry day, with the drum set close and a light draught on the sieves. Peas, linseed, mustard and other crops that ripen irregularly are generally better windrowed. Table IV shows the generally preferred methods for handling various types of seed crops by combine. Further information may be obtained from *Bulletin No. 130* of the Ministry of Agriculture.

ADJUSTMENTS. Leaving a long stubble eases the work of the combine in many ways. On some combines it is shaker capacity that limits threshing capacity, and it is losses from the shakers which become serious if an attempt is made to deal with too much straw. With upstanding crops it is generally advisable to keep well above any weeds or undersown clover seeds.

On many combines adjustment of the drum is effected by raising or lowering the drum, and not by adjusting the concave as with the thresher ; but the principles are exactly the same. The speed of the drum must be reduced for peas and beans, and increased for linseed and clover.

At the beginning of combining the drum speed and concave clearance should be set according to the maker's recommendations. If the combine does not thresh cleanly, the first adjustment should be to increase drum speed. More speed is needed for damp and tough crops. If increasing the drum speed is not effective, the concave clearance must be reduced, care being taken to adjust equally on both sides of the machine.

If threshing is clean and grain is being cracked, the first step should be to widen the concave clearance. After this has been done, if damage to the grain continues, drum speed should be reduced.

Adjustment of the fan needs careful attention. The speed is usually kept constant, and the draught regulated by means of a shutter. Too much draught blows the grain out of the back, while too little fails to keep the screens clear. It is advisable to keep the last quarter of the screen clear on most machines.

The fan blast must be varied in conjunction with adjustment of the sieve opening, and the general principle to be observed is to employ a strong blast and a large sieve opening rather than a weak blast and small sieve opening. Too close a setting of the grain sieve will cause an excessive amount of material

TABLE IV.—*Harvesting Seed Crops by Combine*

Crop	Generally preferred method of harvesting by combine	Notes
Cocksfoot and Fescues	Use binder; stook and thresh from stook. Rarely direct combine.	Medium drum speed and concave clearance. Very light wind. If windrowed, difficult to pick up. If direct-combined, risk of serious loss by shedding, and trouble due to green seeds.
Ryegrass	Windrow and pick-up, or use binder and stook.	High drum speed, small concave clearance, and light wind. Keep windrows small and turn gently. Light, standing crops can be direct combined.
White Clover	Windrow and pick-up.	Filler plates needed. High drum speed, very small concave clearance and light wind. Clover sieves needed. Thresh when perfectly dry. Watch for splitting of seed. Can occasionally be combined direct.
Red Clover	Direct combine unless heavy, leafy crop.	Adjustments as for white clover. Windrow if a heavy, tangled crop.
Sainfoin	Windrow and pick-up. Sometimes direct combined.	" Seed " is enclosed in a husk which has to be milled later.
Trefoil	Windrow and pick-up.	Not generally desirable to attempt to thresh the seed from the pod.
Timothy	Binder and stook. Thresh from stook after several weeks.	Adjustments as for clovers. Threshing is difficult, and re-threshing is sometimes worth while.
Mustard, Rape, Swede, Turnip	Windrow and pick-up.	Medium drum speed and concave clearance. Light wind. Avoid splitting seed.
Linseed	Direct combine if clean.	Medium drum speed, small concave setting, medium wind. Leave till dead ripe. Watch for leaks and stop up with plasticine. If much green rubbish use binder, and stook or windrow.
Sugar beet and Mangold	Binder and stook. Thresh from stook.	Low drum speed, wide concave clearance, medium wind. Very stout and heavy crops much better cut by hand.

to be returned to the drum, and this may result in the grain being cracked.

Losses from combines may be serious unless a constant watch is kept for signs of unsatisfactory threshing. Conditions change from hour to hour, and it is often necessary to vary adjustments several times a day.

The moisture content of grain threshed with the combine varies greatly, according to the state of maturity of the crop and the conditions in which it is harvested. It cannot be stored in bulk with safety if its moisture content exceeds about 14 per cent., and it is often necessary to remove some of the moisture before storage. The construction and operation of grain driers are dealt with in Chapter Nineteen.

Handling of the Straw. There is great diversity of practice in the handling of straw left by the combine. Where it is to be ploughed in the combine may be fitted with a straw spreader. Even distribution of the straw over the stubble greatly facilitates subsequent tillage operations.

On most farms the straw is needed for feeding or for litter and must be collected. Sweeps, hayloaders or balers may be employed, but by far the most popular method is to use a pick-up baler. Some combines may be fitted with a straw deflector which puts two windrows into one, and in light crops this reduces the time spent on baling.

One imported German combine (the Claas) is equipped with a low-density straw trusser which trusses the straw as it comes from the shakers. This fitting is particularly useful where long straw is needed for such purposes as potato clamping.

Much of the saving of time and cost brought about by using a combine rather than a binder may be nullified by inefficient straw handling methods. Some farmers who are concerned to get the straw collected in good condition find it most satisfactory to use a binder, and cart direct to a stationary thresher.

Choice of Crop Varieties. Farmers who use combine harvesters need to take careful account of harvesting characteristics in their choice of variety. The advantage of growing short-strawed crops which are resistant to lodging is now becoming widely recognized, and it is fortunate that most of the short stiff-strawed wheat varieties are also early in ripening. The factor of earliness in ripening is often given insufficient weight in selection of varieties. Earliness is a good feature in itself, since there is more chance of getting long drying days in July and early August than in late September. Earliness is also useful in enabling the harvest season to be spread by the earlier start. Autumn-sown Pioneer barley and Picton oats are examples of very early-maturing varieties, but reference should be made to a series of farmers' leaflets, published by the National Institute of Agricultural Botany and to *Cereal Varieties in Great Britain*, by R. A. Peachey (Crosby Lockwood) for details of suitable crop varieties for various soil and climatic conditions.

There is little to be said for growing crops for combining which are known to mature very late in the year.

Combine Harvester or Binder. Combining has many advantages over use of a binder, especially on large farms. Its chief advantages are reduction of the labour needed for the harvest, elimination of the tedious and costly winter job of threshing, and the fact that with efficient management and a drier available when required, there is less risk of loss due to unfavourable weather. Combining also enables more corn to be gathered from laid crops if devices such as a pick-up reel are used. Nevertheless, there is still a place for the binder on many farms where long straw is valued, and as has been shown already, there are now devices such as hydraulic push-off stackers which may completely alter the appearance of the sheaf-handling problem. Moreover, there is no real reason why stationary threshing should be so slow and costly in labour as it usually is. The binder method is more likely to give a sample of corn that will store safely without drying, and stationary threshing gets most of the weed seeds out of the sample without scattering them over the fields. There are, therefore, many reasons why the occupiers of small farms should interest themselves in possibilities of improving sheaf-handling systems as well as in combine harvesters. The combine, in solving one problem, often introduces two—grain storage and straw handling—which are just as troublesome. It seems likely that any further great extension of the combine method on to the small farms of Britain will require either an increase in contract work, or co-operative use of machinery, or some new cheap combine which either bundles the straw or facilitates its collection by machines which have several other uses.

REFERENCES

(23) " Harvesting by Combine and Binder." Cambridge Univ. Dept. Agric., Farmers' Bulletin No. 9.

(24) Hutt, A. C. *Combine Harvesting and Grain Drying*. London & Counties Coke Association. London.

COMBINE HARVESTERS

The territory in which the combine harvester is used has spread from the rolling wheatfields of the Pacific Coast states to the fields of practically every section of the country. Each year finds this cost-reducing, labor-saving machine proving its value to farmers in new regions and in new crops heretofore thought impractical to combine. In recent years, the small combine, serving the individual farmer having comparatively small acreage, has made the harvesting of small grain and many seed crops a family affair. During the immediate post-war period the self-propelled combine proved its value as still a better means of harvesting large acreages having high yields with less man power and at still a further reduction of cost per bushel.

The combine, in many cases, effects a saving of from fifteen to twenty cents per bushel in harvesting costs. It

Figure 179—A small combine harvesting a fine crop.

displaces the binder, hand shocking, pitching, and threshing. In one operation, the grain is cut and threshed, the cleaned grain elevated into a storage tank, and the straw scattered on the field to be plowed under for humus.

Combines may be divided into two general classifications —pull-type (also called tractor-drawn) and self-propelled.

Pull-Type. The pull-type combine, as the name implies, is drawn by a tractor. The smaller combine, commonly referred to as a one-man machine, derives power for its operation from the power take-off shaft of the tractor, though an auxiliary engine can be furnished if desired. The larger combine has an auxiliary engine which operates the combine mechanism, leaving the weight of the machine the only load for the tractor to pull.

The size of the combine and its detail design is largely dependent upon the width of cut that the machine is expected to handle and the location of the platform or header in relation to the cylinder.

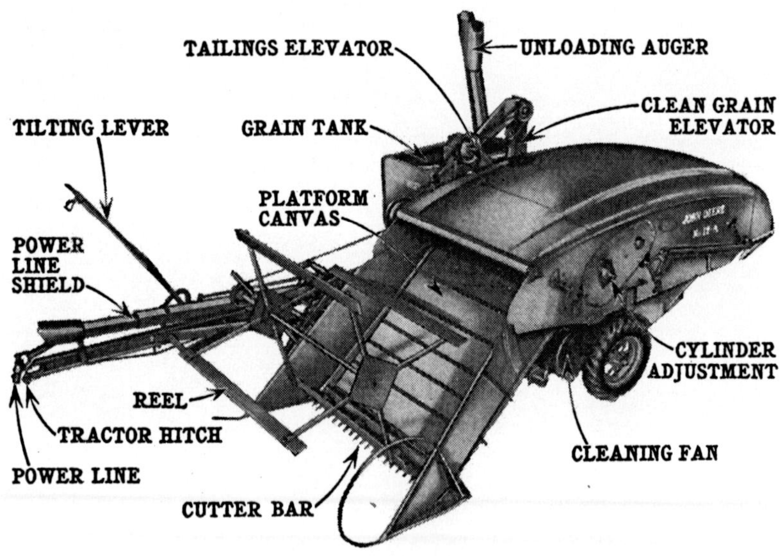

Figure 180—The one-man combine with principal parts indicated.

Straight-through, full-width design is limited to the small combine, usually having not more than a 6-foot cut. Straight-through means handling of the grain and straw in a straight line from cutter bar on through the machine—there are no turns. Full width means that the threshing and separating units are the same width as the cutter bar. A typical straight-through, full-width combine is illustrated in Figures 179 and 180.

Larger combines obviously cannot be made full width but usually employ straight-through design from the feeding unit on back. Because of the large cut the platform is usually to the right or left of the separator and, therefore, the cut grain must make a corner from the platform to the threshing unit. A combine of this type is illustrated in Figure 181.

The combine illustrated in Figure 181 is operated by only

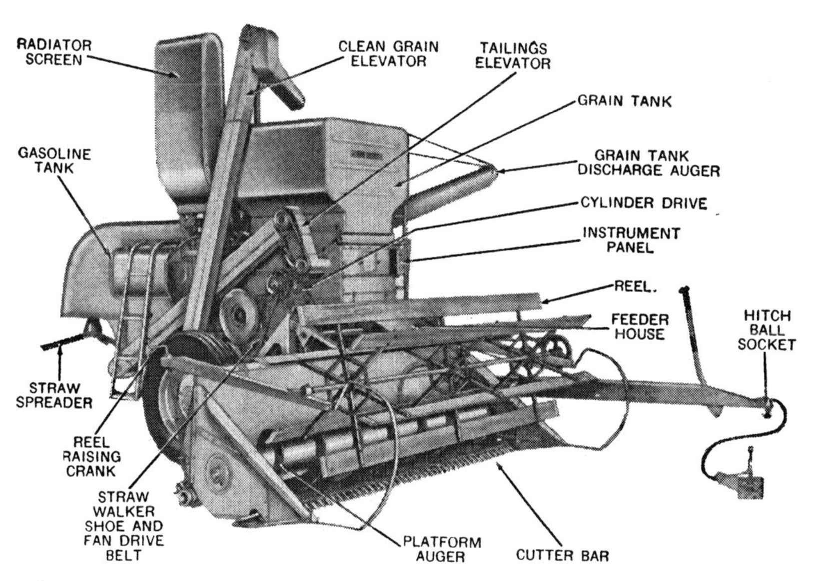

Figure 181—This 12-foot combine is operated by one man, the tractor operator.

one man even though it is a large combine having a 12-foot cut. All controls necessary for operating the combine are located within easy reach of the tractor operator. Formerly, combines of this size required at least two men—one on the combine and one driving the tractor. With this combine, one man and one or two additional men to haul the cleaned grain away are the combine crew. When contrasting the size of this crew with the crew required to operate binders and a stationary threshing outfit, it is apparent that a great saving in labor cost can be made with a combine.

Most combines are designed to operate over fairly level

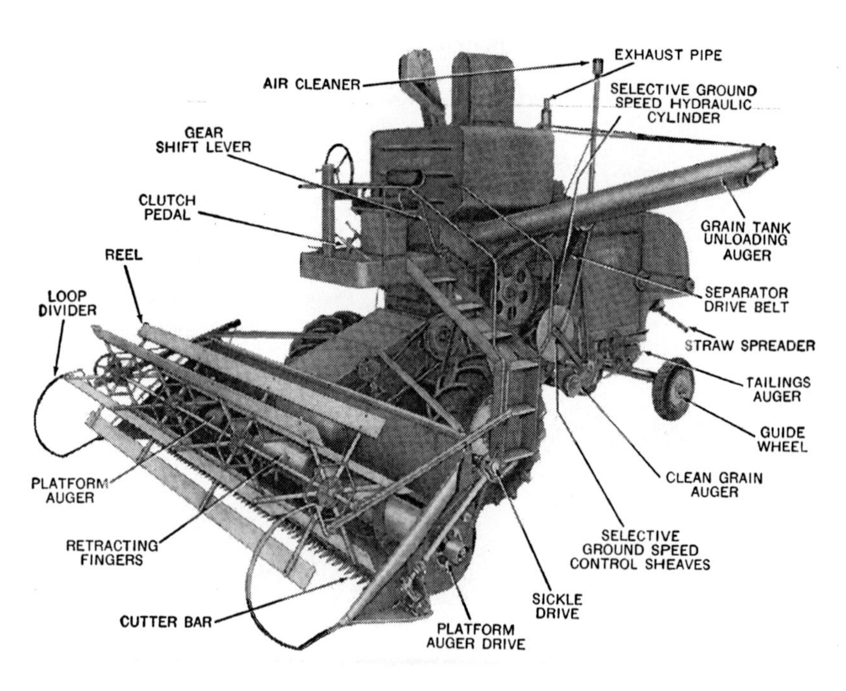

Figure 182—Some important units of the self-propelled combine are indicated in this view.

Figure 183—Cross-sectional view of straight-through combine showing how grain passes through the machine. See description below.

This cutaway view of the straight-through combine shows how the grain and straw are handled from the cutter bar straight through the machine.

The four-slat, ground-driven reel, "A," divides the grain and holds it to the cutter bar, "B." The cut grain is elevated by platform canvas, "C," which, together with feeder, "D," delivers grain in a thin, even stream to the rasp-bar cylinder, "E."

As the grain travels between cylinder, "E," and concave and perforated grate, "F," and back against beater, "G," behind cylinder, the greater part of separation takes place. The grain falls through perforated grate to shoe pan, "K," and is moved back to shoe chaffer, "L." Beater, "G," deflects grain down through the chaffer section at the front end of the straw rack, and passes the straw onto full-width straw rack, "I." During its outward movement, the remaining grain falls through cells in rack onto grain conveyor, "J," and is delivered back to shoe pan, "K," which moves it to front end of chaffer. Straw is then tossed out on the ground.

A blast of air from fan, "N," is directed by deflector, "O," against shoe chaffer, "L," and shoe sieve, "M." This blast, with the aid of chaffer and sieve agitation, blows chaff away and moves the tailings to tailings auger, "P." This auger carries them to tailings elevator, "Q," which conveys them to auger, "R," where they are delivered to the center of the cylinder for re-threshing.

Clean grain, after dropping through shoe chaffer, "L," and shoe sieve, "M," is carried by clean grain auger, "S," to elevator, "T," on opposite side of combine and elevated into grain tank.

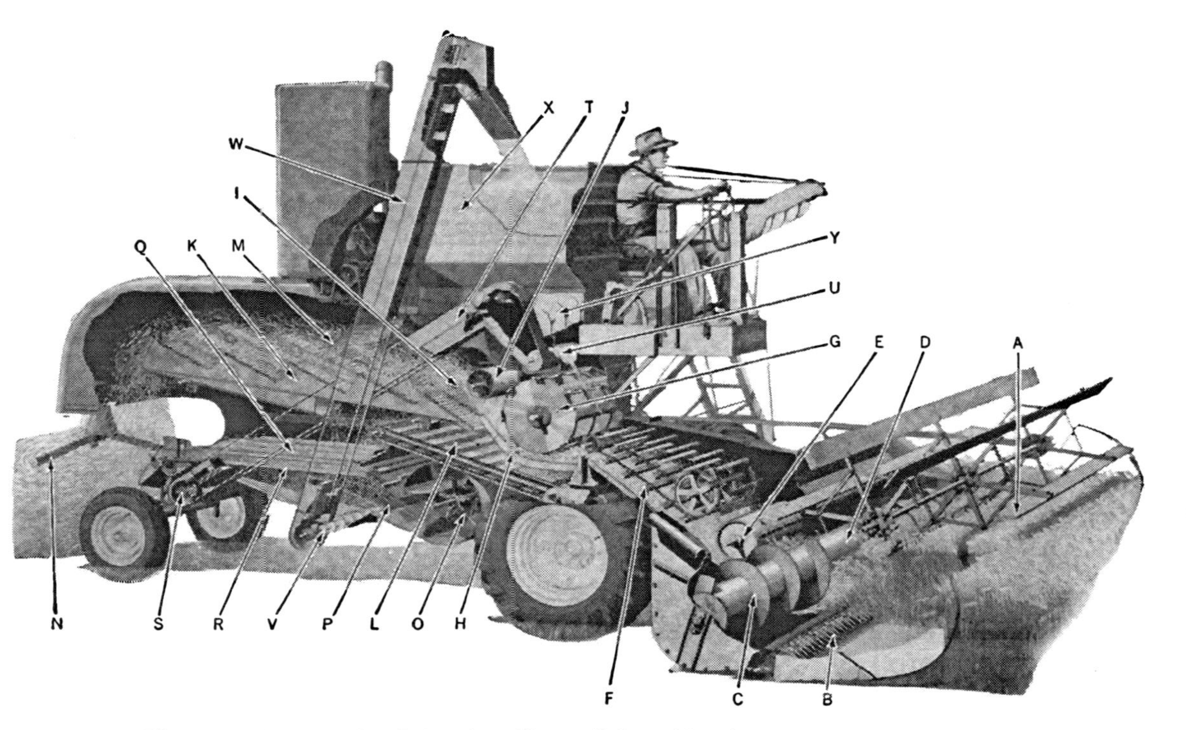

Figure 184—Cross-sectional view of a self-propelled combine showing the important parts. Explanation of the progress of the grain through the machine appears below.

This cut-away view of the modern self-propelled combine shows how the grain and straw are handled from the cutter bar on through the machine.

The power-driven reel, "A," divides the standing grain and holds it to the cutter bar, "B," until cut. The continuous auger, "C," carries the grain from both ends of the platform to the center of the auger, "D." Retracting fingers in beater, "D," take the material and feed it positively to feeder beater, "E," which, in turn, feeds it to the floating undershot feeder conveyor chain, "F." The feeder conveyor chain, "F," delivers the grain in a steady, positive stream to the extra-large, clean-threshing rasp-bar cylinder, "G."

As the grain travels between the cylinder, "G," and concave grate, "H," over grate fingers, "I," and back against the separating cylinder, "J," behind cylinder, the greater part of separation takes place. Separating cylinder, "J," strips straw from the cylinder, deflects grain through the grate fingers, "I," and passes the straw onto straw walkers, "K." Most of the grain falls through concave grate, "H," and fingers, "I," onto conveyor, "L," below cylinder.

Straw and remaining loose grain are passed along to straw walkers, "K." Curtain, "M," keeps grain from being thrown over. Straw is agitated by straw walkers, "K," on its outward movement, and the remaining grain falls through openings in walkers and flows back to shoe chaffer through grain return pans. Straw is then tossed onto spreader, "N."

The grain and chaff from conveyor, "L," are delivered to the cleaning shoe, "Q" and "R." A blast of air from undershot fan, "O," through adjustable windboards, "P," is directed against chaffer, "Q," and lower sieve, "R." This, with aid of sieve agitation, blows chaff away and moves tailings to tailings auger, "S." This auger carries them to tailings elevator, "T," which conveys them through cross auger, "U," to the center of the cylinder, "G," for re-threshing.

Clean grain, after dropping through chaffer, "Q," and sieve, "R," is carried by clean grain auger, "V," to elevator, "W," which delivers it to grain tank, "X." "Y" is grain tank unloading auger.

fields. Where grain is grown on hills or mountain sides as in the Pacific Northwest, a special hillside combine is used. The hillside combine shown in Figure 190 has a special leveling device to keep the separator level whether operating on uphill or downhill grades.

Self-Propelled. The self-propelled combine provides its own motive or propelling power as well as power for cutting, threshing, separating and cleaning the grain. The tractor is released for completing work on the farm; harvest fuel costs are cut about one-third. The self-propelled combine is one-man operated, and the operator, seated high on the machine, has a clear, direct view of his work with all controls conveniently located for easy adaptability to changing field conditions. In fundamental design, it is of the "straight-through" type with the cut grain being delivered to the center of the platform and then up into the threshing unit. No grain is run down when opening fields and the operator may start combining anywhere leaving green spots in the field until later.

Some self-propelled combines as the one illustrated in Figure 182, permit working at any combining speed from a mere crawl on up. This accurately adapts the speed of travel to the capacity of the machine. Increasing or decreasing speed within the three forward transmission gear ranges is accomplished without stopping to shift gears.

The self-propelled combine is especially adapted to the grain harvest on large acreages of the Great Plains and, with special equipment, to meet conditions as they exist in the rice harvest. It is being universally accepted in all sections of the country and in all combineable crops where harvest time is at a minimum and where there is sufficient acreage to warrant a self-propelled combine. The 12-foot self-propelled combine can easily harvest 40 to 50 acres a day.

The small combine shown in Figure 180, is largely designed for the small or medium-sized farm where diversified farming is practiced. It has been used successfully in harvesting many different crops, including soybeans, in addition to all small grains and many seed crops. With this machine, the farmer with an average acreage of small grains can handle his harvest at lowest possible costs.

Principle of Operation. The combine performs four major operations—it cuts the grain, threshes or beats the kernels from the heads, separates the kernels from the straw, and cleans the grain, removing dirt and chaff before the grain is elevated into the storage tank.

The cutting unit which consists of the cutter bar, reel, platform, and feed mechanism operates in much the same manner as a binder, with the exception that it usually cuts higher taking only as much straw as necessary to get all the heads and delivers the heads and the straw into the threshing unit. Manual, hydraulic, or electric lifting mechanisms are used for raising or lowering the platform to meet varying crop conditions. Adjustment and speed of reel are very important in getting the grain into the platform without loss.

The rubbing or flailing action of the cylinder and concave grate in the threshing unit loosens the grain from the heads. Correct cylinder speed and cylinder and concave clearance are all important for good threshing and vary according to the crop. Cylinder speed should be as low as possible and cylinder and concave clearance as high as possible for thorough threshing without excessive cracking of grain.

The separating unit agitates the straw after it comes from the threshing unit, shaking out the loose grain and delivering it to the cleaning unit. Most modern combines separate up to 90% of the grain at the threshing unit, preventing remixing of the grain with straw, thereby making separation of the remaining loose grain comparatively easy. Usually, a separating cylinder or winged beater is placed directly behind the threshing cylinder to slow down the speed of the straw as it comes a mile a minute from the threshing unit. This allows the straw agitating unit to do a thorough job of separating.

Thorough cleaning plays an important part as to the price the farmer gets for his grain. The cleaning unit, consisting of a fan and the shoe containing the cleaning sieves, must remove all foreign material in the grain. The fan blast together with the shaking action of the shoe should keep the chaff lifted slightly off the sieves moving it to the rear and out of machine. Effective cleaning requires the intelligent use of a number of adjustments built into the cleaning unit. Unthreshed heads, commonly called tailings, which pass over the sieves are returned to threshing unit for rethreshing.

Windrowing Method. In many conditions, it is desirable to cut the grain with a windrower and thresh it later with the regular combine equipped with pickup attachment. When there are many weeds in the grain, when there is considerable moisture at harvest time, or when the crop ripens unevenly, this method of combine harvesting is used to advantage.

Figure 185—Harvesting a fine crop of wheat with the self-propelled combine.

The windrower, shown in Fig. 187, consists of the usual cutting-unit platform, sickle, canvases, and reel. These parts are driven either by a power drive shaft from the tractor or through a ground drive. An opening at the inner end of the platform permits the cut grain to be laid on the stubble in windrow form.

When the grain is properly cured or when the moisture content is sufficiently low, a special windrow pickup platform is attached to the combine, or the pickup unit is attached to the regular combine platform. Its function is to elevate the grain onto the combine platform and, from this point on, the threshing, separating, and cleaning processes are the same as described for the regular combine. Pickup attachments are illustrated in Figs. 188 and 189.

The windrow method of combine harvesting has extended the boundaries within which the combine may be used. Many sections where weeds or rainfall have delayed the introduction of the combine are now using the windrow

Figure 186—Here the self-propelled combine is at work in conditions typical of the rice harvest.

method with remarkable success.

Operation and Care. All details of the operation, care, and repair of combines will not be given in this text because the actual servicing of combines will vary widely with the type and manufacturer's design. It is wise, therefore, to follow carefully the operating instructions furnished by manufacturers with the machines they sell. The operator should familiarize himself with these servicing and operating instructions so that he can operate his equipment with greatest efficiency.

Certain essentials for successful operation are stressed by all manufacturers, however, and some of these are mentioned here.

The maximum saving of grain and the quality of work done in all conditions depend very largely upon the operator's making the best use of the adjustments provided for varying conditions. The grain in the tank, the tailings, the straw coming over the straw walkers or racks, and the material going over the shoe reveal the quality of work being done and indicate what adjustments are necessary.

The tractor or combine operator should vary the travel

Figure 187—Windrower used for cutting and windrowing the grain; grain is later picked up with the pickup attachment used on a regular combine, and threshed.

speed to meet conditions. As he approaches a very heavy, down, or tangled condition, he should slow down to give the combine a chance to do a thorough, clean job of separating. In some conditions, it may be advisable to cut less than a full swath, giving the combine every opportunity to do good work. By listening constantly to the sound of the engine, the tractor operator can tell approximately how fast he should travel.

The combine operator should not only adjust his machine to hold threshing losses to a minimum, but he should also adjust and operate the platform and reel to reduce cutting losses. He should watch for stones and other foreign material that the platform may gather, and stop his machine before such obstructions reach the cylinder and cause damage.

Safety first should always be the rule when working around a combine. Never attempt to make repairs while the machine is running. Be careful when working around belts and chains. The great majority of accidents around combines results from carelessness.

Proper lubrication is of first importance. The large number of bearings in a combine necessitate careful attention to regular and thorough oiling. The combines shown are provided

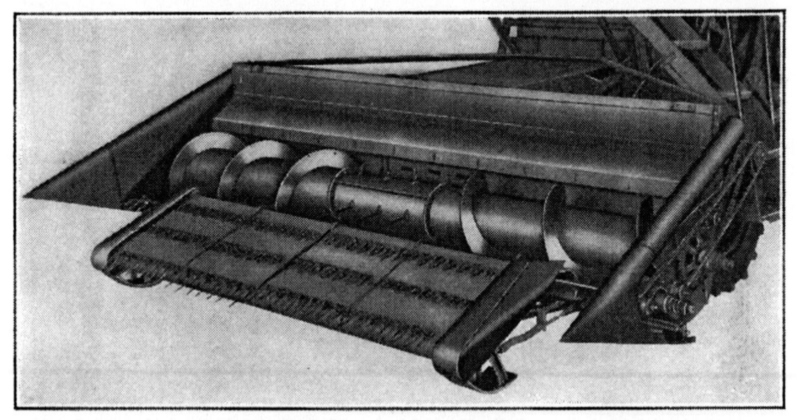

Figure 188—Pickup attachment for combine attached to the short pickup platform. The windrowed grain is elevated onto the platform which carries it into the combine in the regular way.

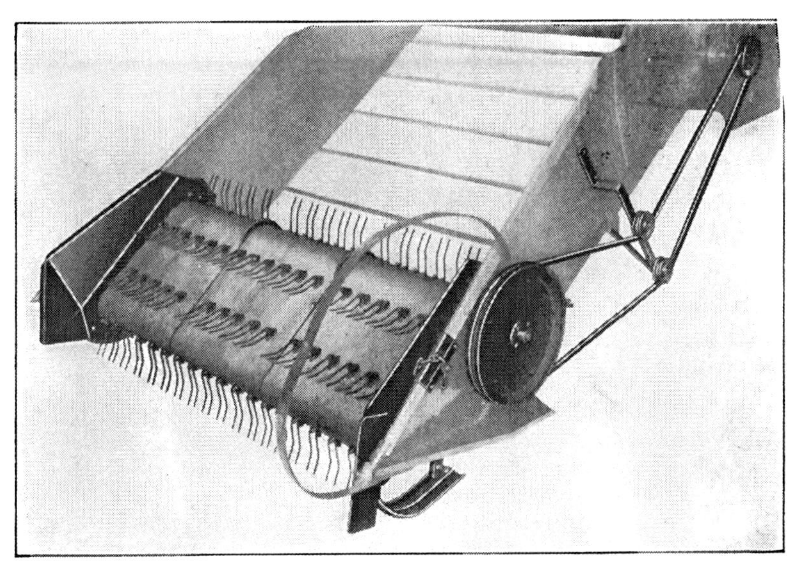

Figure 189—Pickup attachment shown in place on a one-man straight-through combine.

with a high-pressure grease-gun lubrication system which greatly facilitates proper lubrication.

Figure 190—A hillside combine at work in the far western wheat territory.

The upkeep expense on a combine will be greatly reduced if all bolts are kept tight, and belts, canvases, and chains operated at correct tension. Regular inspection of the entire machine will save delays and reduce operating costs to a minimum.

At the end of each season it is important that all dust and chaff be cleaned from the inside and outside of the machine. If not removed, such material will gather moisture and cause steel parts to rust and wood to swell or rot. It will pay the owner of a combine to overhaul and clean the machine thoroughly at the close of each season.

Questions

1. *Name two general types of combines.*
2. *Describe the principle of the combine.*
3. *What are its advantages over other methods of harvesting small grains?*
4. *What are some of the important points to remember in operating a combine? In storing it?*
5. *Describe the windrow method of combining and the machines used.*
6. *How may the tractor operator aid in doing a clean job of harvesting?*
7. *Why is "Safety First" a good motto for combine operators?*

Figure 191—An auxiliary engine furnishes power for operating the small combine shown here.

Lightning Source UK Ltd.
Milton Keynes UK
UKOW04f1103291215

265472UK00001B/171/P